LILIANA THE SUPERHERO
SAVES THE PLANET FROM GLOBAL WARMING

Valeria E. Flores Reinoso

And

Joanne Rolon, Ph.D.

Illustrated by:

Valeria E. Flores Reinoso

ISBN-13: 978-1507709016
ISBN-10: 1507709013

DEDICATION

To all the superheroes in our lives.

Once upon a time not too long ago, there was a little girl named Liliana. She was reading a book with her grandmother when she realized that in all the books she read, the superhero was always a boy. Liliana asked her grandmother why the superheroes were always boys. The grandmother explained to her that there are different kinds of superheroes, and that a superhero is anyone who is willing to do extraordinary things to help others.

The grandmother also told her that there are many girls who are superheroes even if there aren't many books about them. Liliana then told her grandmother that she wanted to be a superhero. The grandmother smiled at her and told her that if she wanted to be a superhero, then all she had to do was be good to everyone, and help others as much as she could.

The next day on her way to school,
Liliana thought of ways she could help others.

Should I help the firemen put out fires?

No, I have to grow up to fit into their uniforms.

Should I travel around the world bringing food to people in need?

No, I'm not allowed to travel alone.

Oh, I know, I'll rescue all the little puppies that don't have a home.

No, my mom won't let me have that many pets. But then, what can I do?

The next day while Liliana was at school, her teacher was talking about global warming.
Global warming? What is that? – Liliana thought.

She paid attention to what her teacher was saying, and then she understood; the planet needed people much like herself to stop global warming and save everyone. Unless the planet stops heating up, there will no longer be a planet to call home.
I don't want to lose my planet - Liliana desperately thought. But what can I do?

She stayed after class and asked her teacher about how she could help.

There are many things you can do, the teacher happily said. You can start by reducing pollution and creating less waste.
How can I create less waste? - asked the little girl.

You can donate your used clothes and toys instead of throwing them away.

You can help your parents recycle things like cans,
bottles, plastic, and paper.

You can also help by using less electricity.

Every time you leave a room, remember to turn off the lights.

Just by doing these simple things you can help stop global warming.

Liliana was very happy and excited to find something she could do. She wanted to save the planet, and now she knew exactly what to do.

When she got home, Liliana explained to her family all the things she had learned.

Together they started a new way of life, reducing their pollution by creating less waste.

The next day Liliana took her bicycle and went around her neighborhood. She told all her friends about the small changes they could make to help save the planet.

Very soon, everyone in her town knew about Liliana.
They followed her advice and decided to help.

Even the mayor of the city heard about the little girl. She was invited to talk on T.V. She told the whole world about global warming and how they could help.
The mayor thanked her for helping save the world. He called her a superhero and gave her a little green cape.

When she went home, her grandmother told her; I saw my favorite superhero on T.V. today.
Who? Liliana asked. You – the grandmother said.
You are my favorite superhero.

THE END

www.ingramcontent.com/pod-product-compliance
Lightning Source LLC
Chambersburg PA
CBHW041614180526
45159CB00002BC/851